Visit Susie and her Animal Friends at
KidsKyngdom.com

Visit Susie and her Animal Friends at
KidsKyngdom.com

For my Lovely Niece

Emily

Thank You for Showing us how to Stretch our Wings and Soar with Delight

Visit Susie and her Animal Friends at
KidsKyngdom.com

Thank you for supporting Kids Kyngdom

Get Your **FREE** audio book at KidsKyngdom.com

Hi **Kyngs**

Welcome to the Namib desert

Visit Susie and her Animal Friends at
KidsKyngdom.com

It is Huge!

Visit Susie and her Animal Friends at
KidsKyngdom.com

It is Right Next to

The Atlantic Ocean

Visit Susie and her Animal Friends at
KidsKyngdom.com

It can be Bitter Cold and Super Hot

In the **SAME** Day

Visit Susie and her Animal Friends at
KidsKyngdom.com

Very Few Humans

Live Here

Visit Susie and her Animal Friends at
KidsKyngdom.com

Desert Elephants call this place

HOME

Visit Susie and her Animal Friends at
KidsKyngdom.com

Visit Susie and her Animal Friends at
KidsKyngdom.com

Visit Susie and her Animal Friends at
KidsKyngdom.com

Desert Elephants are **Skinny** and much **Taller**

than other Elephants

Visit Susie and her Animal Friends at
KidsKyngdom.com

They have
VERY LARGE

FEET

Large feet help desert elephants

Move well in DEEP Sand

Visit Susie and her Animal Friends at
KidsKyngdom.com

Because it is **SO** Hot Desert Elephants Travel at Night

When it is **Nice and Cool**

Visit Susie and her Animal Friends at
KidsKyngdom.com

Spraying Sand on Their Body

Is One Way to Stay Cool

Visit Susie and her Animal Friends at
KidsKyngdom.com

They Spray with Really LONG

TRUNKS

Visit Susie and her Animal Friends at
KidsKyngdom.com

They Dig DEEP in the SAND with their Trunks

To Find Water

Desert Elephants can Live without Water

for 3 Whole Days

I am Getting Thirsty

Are YOU **Thirsty** Bash?

Visit Susie and her Animal Friends at
KidsKyngdom.com

I Remember You Drink Water

Through Your Skin

Visit Susie and her Animal Friends at
KidsKyngdom.com

Did you know that Desert Elephants do NOT Sweat

They Fan their Bodies with their Ears

Visit Susie and her Animal Friends at
KidsKyngdom.com

Desert Elephants Eat Plants

For 20 Hours Every Day

Visit Susie and her Animal Friends at
KidsKyngdom.com

They eat Bushes, Grasses

and **Herbs**

Not this Herb

Visit Susie and her Animal Friends at
KidsKyngdom.com

This Herb

Visit Susie and her Animal Friends at
KidsKyngdom.com

Adult Male Elephants

are Called **BULL** Elephants

Visit Susie and her Animal Friends at
KidsKyngdom.com

Female Elephants travel in small groups

With their Young

Visit Susie and her Animal Friends at
KidsKyngdom.com

I Love visiting animals

with you Bash

Visit Susie and her Animal Friends at
KidsKyngdom.com

Hey Did you know that Hathi means "Elephant"

In the Hindi Language

Visit Susie and her Animal Friends at
KidsKyngdom.com

That's Right!

Visit Susie and her Animal Friends at
KidsKyngdom.com

Hathi also means RARE

Visit Susie and her Animal Friends at
KidsKyngdom.com

Let's take Care of our RARE Desert Elephants

Visit Susie and her Animal Friends at
KidsKyngdom.com

How about we visit another AMAZING Animal?

Visit Susie and her Animal Friends at
KidsKyngdom.com

Join us on our next visit Kyngs

When we visit a **KOALA**

Visit Susie and her Animal Friends at
KidsKyngdom.com

Did you Love
visiting Elephants
with Susie and Bash?

Please Share Your Review

Visit Susie and her Animal Friends at

KidsKyngdom.com

What is Sammie Kyng's Favorite Animal?

Sammie Kyng's favorite animal used to be an Elephant.

Now it's a **Koala!**

If you have a favorite animal you would like Doctor Susie to visit, just send a request to Kidskyngdom.com or Instagram.

KidsKyngdom.Com

Instagram.com/Kids_Kyngdom

Visit Susie and her Animal Friends at
KidsKyngdom.com

ISBN: 978-1-959501-09-1

Published by: Kyngdom, LLC

Visit Susie and her Animal Friends at
KidsKyngdom.com

Do You Know What Name all Baby Koalas are Born with?

Bobby
Cindy
Jane
Robin
Danny
Zeke

Get some Answers with Susie and Bash

<u>**Visit** KidsKyngdom.com</u>

www.ingramcontent.com/pod-product-compliance
Lightning Source LLC
Chambersburg PA
CBHW051347290326
41933CB00042B/3323
* 9 7 8 1 9 5 9 5 0 1 0 9 1 *